COLORING BOOK

Hello friend!

Welcome to a world of relaxation, creativity, and self-discovery! Our coloring book is a delightful escape into a world of intricate designs and artistic expression.

Within these pages, you'll find intricate designs and patterns designed to captivate your imagination and provide a soothing escape from the demands of daily life.

Each page is a canvas awaiting your personal touch, an opportunity to infuse vibrant hues into beautifully crafted patterns. Immerse yourself in the art of coloring and let the stress melt away.

HAPPY COLORING!

Our paper is most suited for colored pencils, crayons, pastels, and alcohol-based markers. To protect upcoming pages, we recommend using a blank paper under the page you're working on.

THIS BOOK BELONGS TO

TEST COLOR PAGE

WRITE OUT YOUR FAVORITE ASPECTS OF THIS BOOK

THANK YOU FOR TRUSTING US BY PURCHASING OUR BOOKS.

Your trust in us means a lot, and we truly hope that you will find joy and satisfaction in coloring our unique designs. If our book meets your expectations, we kindly ask you to leave a positive review as it motivates us to create even better books in the future. Once again, thank you for your support and we hope that our coloring book will bring a little bit of creativity and relaxation into your life.

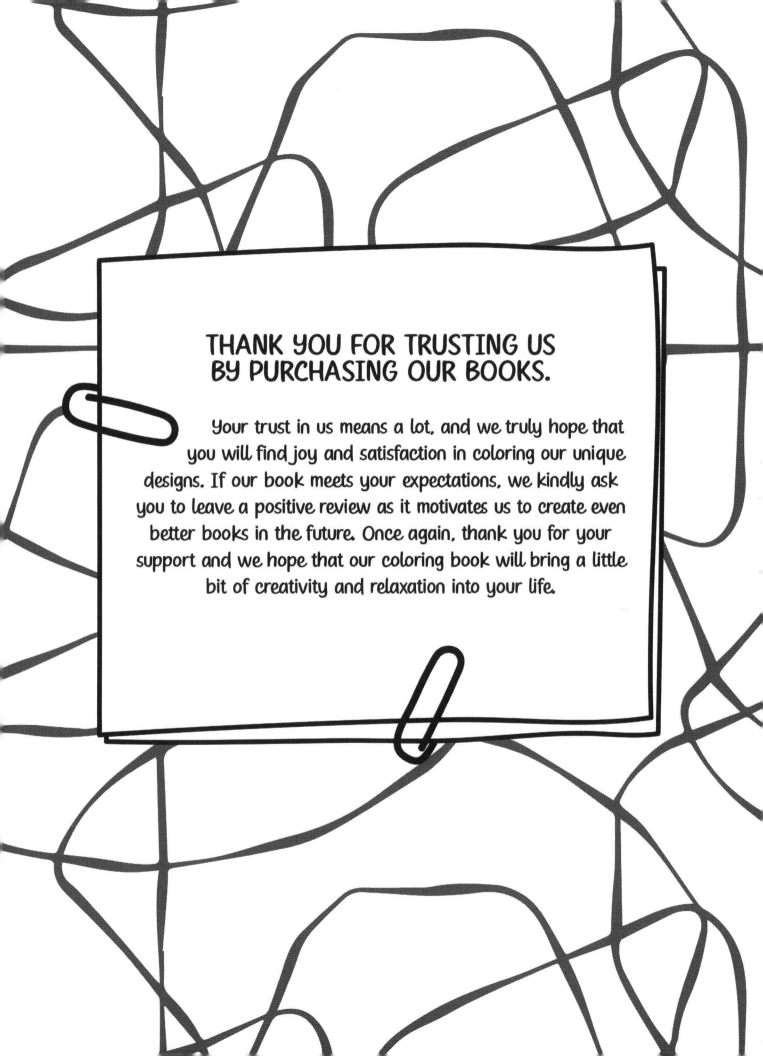

Made in United States
Troutdale, OR
05/01/2024

19573590R00060